DATE DUE	
AUG 1 2 2011		
OCT 15 2011		
DEC 1 8 2012		

The Urbana Free Library

To renew materials call
217-367-4057

Let's Read About Food

Vegetables

by Cynthia Klingel and Robert B. Noyed
photographs by Gregg Andersen

Reading consultant: Cecilia Minden-Cupp, Ph.D.,
Adjunct Professor, College of Continuing and Professional Studies, University of Virginia

WeeklyReader.
EARLY LEARNING LIBRARY

1-03
19.00

For a free color catalog describing
Weekly Reader® Early Learning Library's
list of high-quality books, call 1-800-542-2595
or fax your request to (414) 332-3567.

Library of Congress Cataloging-in-Publication Data available
upon request from publisher. Fax (414) 336-0157 for the
attention of the Publishing Records Department.

ISBN 0-8368-3060-1 (lib. bdg.)
ISBN 0-8368-3149-7 (softcover)

This edition first published in 2002 by
Weekly Reader® Early Learning Library
330 West Olive Street, Suite 100
Milwaukee, WI 53212 USA

An Editorial Directions book
Editors: E. Russell Primm and Emily Dolbear
Art direction, design, and page production: The Design Lab
Photographer: Gregg Andersen
Weekly Reader® Early Learning Library art direction: Tammy Gruenewald
Weekly Reader® Early Learning Library production: Susan Ashley

Printed in the United States of America

1 2 3 4 5 6 7 8 9 06 05 04 03 02

Note to Educators and Parents

As a Reading Specialist I know that books for young children should engage their interest, impart useful information, and motivate them to want to learn more.

Let's Read About Food is a new series of books designed to help children understand the value of good nutrition and eating to stay healthy.

A young child's active mind is engaged by the carefully chosen subjects. The imaginative text works to build young vocabularies. The short, repetitive sentences help children stay focused as they develop their own relationship with reading. The bright, colorful photographs of children enjoying good nutrition habits complement the text with their simplicity and both entertain and encourage young children to want to learn — and read — more.

These books are designed to be used by adults as "read-to" books to share with children to encourage early literacy in the home, school, and library. They are also suitable for more advanced young readers to enjoy on their own.

— *Cecilia Minden-Cupp, Ph.D.,*
Adjunct Professor, College of Continuing and
Professional Studies, University of Virginia

I like to eat
vegetables.
They are good
for me.

We choose from six different kinds of food. We need to eat all six kinds every day to stay healthy.

fats and sweets

milk and cheese

meat

vegetables

fruit

bread and cereal

My body gets
many vitamins
from vegetables.
They help make
me healthy and
strong.

I like vegetables. Vegetables come in many shapes and colors.

Some vegetables are green and leafy. I like spinach and lettuce.

13

Some vegetables taste good cooked. I like cooked string beans and broccoli.

Some vegetables taste good raw. Carrot sticks and celery are a great snack after school.

17

Peas are also vegetables. My favorite soup in the winter is pea soup.

Right now, I am hungry for some corn on the cob!

Glossary

healthy—to be strong and free of illness

snack—a small, light meal

vitamin—one of the substances in food that is needed for good health

For More Information

Fiction Books

Lin, Grace. *The Ugly Vegetables*. Watertown, Mass.:
    Charlesbridge, 1999.
Speed, Toby, and Barry Root. *Brave Potatoes*. New York: Putnam,
    2000.

Nonfiction Books

Barlowe, Dot. *Learning About Vegetables*. New York: Dover,
    2001.
Florian, Douglas. *Vegetable Garden*. New York: Harcourt, 1996.

Web Sites
Phillips Mushroom Place
www.phillipsmushroomplace.com/
For information about mushrooms

The World Carrot Museum
www.carrotmuseum.com
For everything you want to know about carrots

Index

About the Authors

Cynthia Klingel has worked as a high school English teacher and an elementary school teacher. She is currently the curriculum director for a Minnesota school district. Cynthia Klingel lives with her family in Mankato, Minnesota.

Robert B. Noyed started his career as a newspaper reporter. Since then, he has worked in school communications and public relations at the state and national level. Robert B. Noyed lives with his family in Brooklyn Center, Minnesota.